HANDY ARTIFICIAL INTELLIGENCE

(AI) Answer Book

AF807708

By

Hirendra Majumdar, P.E., M. ASCE

TABLE OF CONTENTS

Foreword iv

Chapter 1: Artificial Intelligence And Its Branches 1

Chapter 2: Fundamentals of Machine Learning 29

Chapter 3: Neural Networks and Deep Learning 31

Chapter 4: Natural Language Processing 33

Chapter 5: AI in Real-World Applications 36

Chapter 6: Ethical Challenges in AI 39

Chapter 7: The Future of AI 42

Concluding Thoughts on Artificial Intelligence 44

Appendix A: Data Sources to Train AI Models 45

Appendix B: Exploring AI Hardware, Explainability, and Creativity 48

About the Author 49

FOREWORD

Imagine trying to teach a dog a new trick. At first, it stares at you, confused. But over time, with enough repetition and rewards, it learns. AI, in many ways, learns like this, through trial, error, and patterns. The difference? It can process millions of examples in minutes.

We are living in a time when technology is not just changing our lives—it is reshaping the very fabric of how we live, work, and interact. At the heart of this transformation is Artificial Intelligence (AI), a field that has evolved from science fiction into a powerful force driving innovation across nearly every industry.

This book is a gateway into the fascinating world of AI. Whether you're a curious newcomer, a student, a professional exploring the future of technology, or someone simply wondering how your phone seems to understand you so well. This book will guide you from the basics to the more complex layers of the AI universe.

You'll explore the foundations of AI and discover how machine learning, deep learning, and neural networks function under the hood. But more importantly, you'll see how these technologies are being applied—from self-driving cars and recommendation systems to healthcare diagnostics and creative tools. The goal is not just to inform, but to inspire.

This book is structured around real questions you might ask about AI, each followed by clear, concise answers. Whether you're flipping to a topic or reading cover to cover, this format makes it easier to enter the complex world of AI.

Welcome to a journey into one of the most transformative technologies of our time. Let's begin!

CHAPTER 1

ARTIFICIAL INTELLIGENCE AND ITS BRANCHES

Imagine you hired a super-efficient assistant. At first, you'd teach them how to organize files or answer emails. But then, they start learning on their own, getting better, faster, and even anticipating your needs. That's what AI does for computers.

Artificial Intelligence (AI) is the science of creating machines that can do tasks which normally require human intelligence. These tasks include:

- Learning (acquiring information and rules for using it)
- Reasoning (using rules to reach approximate or definite conclusions)
- Problem-solving
- Understanding natural language (e.g., speech and text)
- Perception (interpreting sensory input, such as vision)
- Decision-making
- Autonomous action (e.g., robotics)

Let's dive in.

A Brief History of Artificial Intelligence

The idea of artificial intelligence has ancient roots. Early myths and philosophical reflections imagined machines that could think or act like humans. Devices like the mechanical pigeon of Archytas hinted at the dream of automation long before computers existed.

In the 1950s, the field took formal shape. The term "artificial intelligence" was introduced at the Dartmouth Workshop in 1956, where pioneers such as Alan Turing, John McCarthy, and Marvin Minsky laid the foundation for modern AI research.

During the 1960s and 1970s, AI development focused on symbolic reasoning and logic-based systems. These early efforts were constrained by limited memory and computing power. As expectations outpaced results, the field entered a period of slowed progress in the late 1970s and 1980s—often called the "AI winter."

Momentum returned in the 1980s with the rise of expert systems. These programs mimicked human decision-making in specialized fields, demonstrating that AI could be useful in real-world applications. This success renewed interest and funding for research.

Since the 2000s, AI has been shaped by breakthroughs in machine learning, especially deep learning. These approaches rely on neural networks to process large datasets, allowing machines to identify patterns and improve performance in tasks such as image recognition, speech processing, and language understanding.

Q1. What are the different types of AI?

In practical terms, AI systems can range from simple rule-based engines to advanced models, such as large language models, self-driving cars, and real-time facial recognition systems. AI can be classified into:

- Narrow AI: Specialized for specific tasks (e.g., voice assistants, spam filters).
- General AI: Hypothetical AI with human-like intelligence across a wide range of tasks. The jack of all trades.
- Superintelligent AI: A theoretical future AI that surpasses human intelligence in all domains.

So far, we only have Narrow AI, while the other two types are in development.

Q2. What are the branches of Artificial Intelligence (AI)?

Branches of AI include:

1. Natural Language Processing
2. Machine Learning
3. Robotics
4. Expert Systems
5. Computer Vision
6. Fuzzy Logic
7. Neural Network
8. Limited Memory
9. Reactive Machines and Deep Learning
10. Self-Aware AI
11. Supervised Learning
12. Pattern Recognition
13. Logical Reasoning
14. Healthcare
15. Planning
16. General AI
17. Reinforced Learning
18. General Intelligence
19. Learning
20. AI in Education
21. Stages of AI

Q3: What is Natural Language Processing (NLP)?

Imagine you're teaching a robot to understand you. First, you show it how to split your sentence into words, then how to figure out what those words mean together. That's what Natural Language Processing (NLP) does.

At its core, NLP involves a mix of linguistics and computer science. Techniques like tokenization (breaking text into smaller parts), parsing (analyzing sentence structure), and machine learning models help computers grasp the nuances of language. Think of it as teaching a machine how to read between the lines.

In terms of a pplications, NLP is the technology behind speech recognition, chatbots, and search engines.

Q4.What is Machine Learning?

Moving on, imagine showing your friend hundreds of photos of different cat breeds until they can tell which is which. Machine learning also learns from examples, not instructions.

- In **supervised learning**, the machine gets the right answers first, like a student using a solved math workbook.
- In **unsupervised learning**, there are no answers, it's like figuring out a puzzle from scratch.
- In **reinforcement learning**, the machine learns through trial and error, like a kid playing a video game.

Machine learning is behind many everyday applications: recommendation systems (like those used by Netflix or Amazon), speech recognition, fraud detection, and even self-driving cars.

Q.5 What is Robotics?

Robotics is the field of engineering and computer science that focuses on designing, building, and programming robots, i.e., machines capable of performing tasks autonomously or with human guidance. Think of a robot as a helpful machine that can vacuum your floor, assist in surgery, or even explore space. Some follow instructions exactly. Others are learning to think on their feet. These robots come in many forms, from industrial robots that assemble cars to humanoid robots that mimic human movements.

At its core, robotics combines multiple disciplines:

- Mechanical engineering for designing physical structures and movement.
- Electrical engineering for powering and controlling robots.
- Computer science and AI for enabling intelligent behavior and decision-making.

As stated above, robots are used in manufacturing, healthcare, space exploration, and even household applications.

Q6. What are Expert Systems?

Think of an expert system like a digital detective. It collects clues (facts), applies rules (logic), and tells you what it concludes, just like an experienced doctor or mechanic.

Expert systems are AI programs designed to solve problems by mimicking how a human expert thinks. They rely on stored knowledge and logical rules to make decisions.

Key components of an expert system include:

- **Knowledge Base:** A comprehensive collection of domain-specific knowledge, including facts, rules, and relationships.
- **Inference Engine:** Applies logical rules to the knowledge base to deduce new information and solve problems.
- **Explanation Facility:** Provides explanations for the conclusions reached by the system.
- **User Interface:** Allows users to interact with the system and input information.

Q7. What is Computer Vision?

Moving on, computer training enables machines to not only scan but also interpret images and video. It leverages algorithms and models, often utilizing machine learning and deep learning, to identify patterns in pixel data. It can label images, detect objects, track movement, or even understand entire scenes.

What it does

Computer vision aims to make computers "see" and understand visual information. This involves extracting meaning from images and videos, identifying objects, and even understanding scenes.

How it works

It uses algorithms, often powered by machine learning and deep learning, to analyze digital images and videos. These algorithms can be trained to recognize patterns, classify objects, and even predict future actions based on visual information.

To sum up, computer vision allows machines to "see" and understand visual information by analyzing patterns in images or videos. It performs tasks like labeling images, detecting objects, and tracking motion. These capabilities power real-world applications such as self-driving cars, medical imaging, security systems, and augmented reality tools.

Q8. What Is Fuzzy Logic?

Fuzzy logic is a branch of artificial intelligence that helps machines handle uncertainty. Unlike traditional logic, which sees things as either true or false, fuzzy logic allows for in-between values. It works with "partial truths," letting a statement be somewhat true and somewhat false at the same time.

How Fuzzy Logic Works

Fuzzy logic systems follow a structured process:

1. **Fuzzification** – Real-world input values are converted into fuzzy sets.

2. **Rule Evaluation** – The system applies "if-then" rules to the fuzzy inputs.

3. **Aggregation** – It combines the results of all the rules.

4. **Defuzzification** – The system converts the fuzzy output back into a clear, usable value.

Key Concepts

Fuzzy Sets

Instead of using sharp boundaries, fuzzy sets allow values to gradually transition from one category to another. For example, the term "hot" doesn't need a fixed temperature—it can describe a range.

Membership Functions

These define how much a value belongs to a fuzzy set. A triangular membership function, for instance, might assign high membership to 85°F and lower membership to 75°F or 95°F.

Fuzzy Rules

These are logic-based "if-then" statements. For example:

If the temperature is high, then the fan speed should be high.

Fuzzy Inference System (FIS)

This system connects all the components. It manages fuzzification, applies the rules, and then returns a final crisp output.

Applications

Fuzzy logic is used across many industries:

- **Control Systems** – Washing machines, air conditioners, and autonomous vehicles
- **Natural Language Processing** – Search engines and text analysis
- **Image Processing** – Segmentation and object recognition

- **Medical Diagnosis** – Assisting in detecting diseases and recommending treatments
- **AI Systems** – Expert systems, decision-making, and pattern recognition

Benefits

- Handles vague or incomplete data
- Mimics human-style reasoning
- Adapts to changing environments
- Improves performance in complex situations

Q9. What Is a Neural Network?

A neural network is a machine learning model designed to work like the human brain. It processes data using layers of interconnected nodes, known as artificial neurons. These networks are trained to recognize patterns and make decisions based on data.

How Neural Networks Work

Neural networks are made up of nodes arranged in layers:

- **Input Layer** – Receives raw data
- **Hidden Layers** – Process the data through weighted connections
- **Output Layer** – Produces the result (like a prediction or classification)

Each node processes information and passes it on to the next layer. As the network trains on more data, it adjusts the connections (or “weights”) between nodes to improve accuracy.

Key Features

- **Inspired by the Brain** – The structure mimics how biological neurons transmit signals
- **Part of Machine Learning** – Neural networks are a core tool in

machine learning

- **Used in Deep Learning** – Many deep learning models are built using neural networks to extract features from large datasets

Applications

Neural networks are used in many AI applications:

- Image and speech recognition
- Language translation and processing
- Fraud detection and financial forecasting
- Medical diagnosis and drug discovery

Q10. What Is Limited Memory in AI?

Limited memory AI refers to systems that can temporarily use past data to make better decisions. Unlike reactive AI, which responds only to current inputs, limited memory systems can learn from recent events and use that information to inform future actions. However, they don't retain this data permanently.

Key Characteristics

- **Temporary Data Use** – These systems access past data for a short time before discarding it.
- **Short-Term Learning** – They use recent patterns to refine their responses.
- **Difference from Reactive AI** – While reactive systems don't "remember" anything, limited memory systems can recall data from recent interactions.

Common Applications

- **Fraud Detection** – Identifying suspicious activity based on recent transaction patterns
- **Spam Filters** – Learning from recent messages to block unwanted content

- **Virtual Assistants** – Responding based on previous user queries
- **Customer Service Chatbots** – Offering tailored support from earlier interactions
- **Self-Driving Cars** – Using nearby vehicle and road data to make split-second decisions
- **Predictive Maintenance** – Forecasting equipment failure from recent performance data

Q11. What Are Reactive Machines and Deep Learning?

Artificial intelligence (AI) includes systems that range from simple responders to complex learners. Two such systems are reactive machines and deep learning models—very different in function and design.

Reactive Machines

Reactive machines are the most basic form of AI. They respond only to current inputs and don't use memory or past experiences to inform decisions.

- No memory or learning capability
- Act only based on present conditions
- Ideal for tasks with fixed rules and limited variability

Example: IBM's Deep Blue, the chess computer, chose moves based solely on the current board, not on prior games or strategies

Reactive AI is still useful in systems like basic spam filters or entry-level recommendation engines.

Deep Learning

Deep learning is a type of machine learning that uses artificial neural networks with many layers. These models can learn patterns from data automatically.

- Use layered neural networks to process data

- Extract features from data without manual coding
- Enable systems to perform image recognition, voice translation, and language generation

Examples: Tools like facial recognition software, voice assistants, and content creation models

Unlike reactive machines, deep learning systems improve over time as they are exposed to more data.

Q12. What Is Self-Aware AI?

Self-aware AI is a theoretical form of artificial intelligence that goes beyond task performance. It would possess an understanding of its own identity, existence, and possibly emotions, much like human consciousness.

Key Characteristics

- **Consciousness** – The ability to be aware of one's own presence and mental state
- **Subjectivity** – A personal, inner experience or point of view
- **Self-Referential Thinking** – Awareness of one's own thoughts, actions, and decisions
- **Potential for Emotions** – The idea that machines could one day simulate or experience emotional states

Challenges and Ethical Considerations

- **Understanding Consciousness** – We don't yet fully understand how human consciousness works, making it difficult to model artificially
- **Creating Subjectivity** – Building machines with inner experience and emotional awareness is an unsolved technical challenge
- **Ethical Questions** – If machines become self-aware, how should they be treated? What moral rights would they have?

Current State of AI

Today's AI systems, including large language models, can process language, make predictions, and generate content. But they do not possess consciousness or awareness.

Some robotic systems show a limited form of self-awareness by monitoring their position, movement, or environment. Recent research is exploring how robots might better adapt and learn through environmental feedback, but this still falls far short of true self-awareness.

The Road Ahead

Self-aware AI remains speculative. It represents a possible future where machines evolve from tools into entities with identity and perception, a shift that would redefine the boundaries between human and machine intelligence.

Q13. What Is Supervised Learning?

Supervised learning is a method in artificial intelligence where machines are trained using labeled data. This means the training data includes both the input (such as a photo or a sentence) and the correct output (like a label or value). The goal is to help the model learn the connection between them so it can make accurate predictions on new data.

How It Works

Labeled Data

In supervised learning, every example in the dataset includes a known label or target. For instance, an image might be labeled as "cat" or "dog," allowing the system to associate image features with the correct classification.

Predictive Modeling

As the model processes these labeled examples, it adjusts its internal rules to improve accuracy. Once trained, it can analyze new, unlabeled data and predict the correct output.

Types of Tasks

Classification

The model predicts a category.

Example: Determining if an email is spam or not.

Regression

The model predicts a number or continuous value.

Example: Estimating the price of a house based on location and size.

Real-World Applications

Supervised learning is widely used in:

- Spam detection
- Image and speech recognition
- Fraud detection
- Weather forecasting
- Stock market prediction

Comparison with Unsupervised Learning

Unlike supervised learning, unsupervised learning works with data that has no labels. Instead of predicting known outcomes, the model tries to uncover hidden patterns or groupings on its own.

Q14. What Is Pattern Recognition in AI?

Pattern recognition is the ability of AI systems to detect and understand recurring features in data. This includes identifying visual patterns in images, trends in financial data, or sound sequences in speech.

Why It Matters

Pattern recognition is a core function of machine learning. It enables machines to "learn" from examples and make decisions based on what they've seen before. It helps transform raw data into insights by spotting trends or relationships.

Applications

Pattern recognition is used across many industries:

- **Image Processing** – Detecting objects, faces, or features
- **Speech Recognition** – Converting spoken language to text
- **Medical Imaging** – Identifying symptoms or abnormalities
- **Data Analytics** – Spotting trends, fraud, or unusual activity

Techniques Used

Several AI methods power pattern recognition:

- Machine learning algorithms
- Artificial neural networks
- Statistical analysis and probabilistic models

Types of Pattern Recognition

- **Classification** – Assigning data into categories (e.g., spam or not spam)
- **Clustering** – Grouping similar data points based on shared traits

Real-World Example

Microsoft Copilot describes pattern recognition as the ability to spot recurring trends or relationships in data, images, sounds, or behavior. For example:

- AI facial recognition systems detect and compare facial features
- Financial software tracks stock price trends
- The human brain uses it when identifying a familiar song after just a few notes

Pattern recognition bridges the gap between raw data and intelligent decision-making. It helps both humans and machines make sense of the world by recognizing what's familiar, similar, or repeating.

Q15. What Is Logical Reasoning in AI?

Logical reasoning in artificial intelligence refers to the ability of machines to make decisions or draw conclusions using formal logic. These systems rely on rules, facts, and algorithms, similar to how humans think logically.

How It Works

AI utilizes structured representations of knowledge, such as rules, premises, and facts, to draw new conclusions. This is made possible by:

- **Formal Logic Systems** – Such as propositional and first-order logic
- **Algorithms** – That apply logic rules through deduction, induction, or other methods
- **Knowledge Representation** – Using formats like ontologies or graphs so AI can "understand" and reason with information
- **Inference** – Drawing new conclusions from known data

Logical reasoning enables AI systems to solve problems, make predictions, and manage knowledge in a systematic way.

Types of Logical Reasoning

Logical reasoning in AI includes several styles:

Classic Reasoning

- **Deductive Reasoning** – From general rules to specific outcomes

  If all humans are mortal, and Socrates is a human, then Socrates is mortal.

- **Inductive Reasoning** – From observations to generalizations

  The sun rises daily, so it will rise tomorrow.

Explanatory and Comparative Reasoning

- **Abductive Reasoning** – Inferring the best explanation

  The alarm didn't ring—maybe the battery is dead.

- **Analogical Reasoning** – Solving new problems by comparing them to known situations

  If a past issue was fixed with a certain method, that method might work again.

Adaptive Reasoning

- **Common Sense Reasoning** – Using everyday knowledge

  If you drop glass, it will likely break.

- **Monotonic vs. Non-Monotonic** – Whether new information changes previous conclusions
- **Fuzzy Reasoning** – Reasoning under uncertainty using partial truths

Applications in AI

Logical reasoning supports a wide range of AI functions:

- **Natural Language Processing (NLP)** – Understanding sentence structure and meaning
- **Expert Systems** – Simulating decision-making in specialized fields
- **Automated Theorem Proving** – Solving mathematical problems with logic
- **Decision-Making** – In fields like healthcare, robotics, and finance
- **Problem Solving** – Breaking down complex scenarios into logical steps

Real-World Example: Microsoft Copilot

Microsoft Copilot describes logical reasoning in AI as the ability to make decisions using rules and facts. For example:

- **Deductive**: "All birds have feathers. Penguins are birds. So, penguins have feathers."
- **Inductive**: Observing daily behavior to predict future patterns.

This structured reasoning helps AI ensure consistency and reliability in applications like automated planning, language processing, and expert decision systems.

Q16. AI in Healthcare: Transforming the Medical Field

Artificial Intelligence (AI) is transforming nearly every aspect of healthcare, from diagnosing diseases and creating personalized treatments to streamlining hospital operations and supporting medical research.

1. Diagnostics and Treatment Advancements

Improved Accuracy

AI analyzes medical scans, lab results, and patient histories to detect diseases earlier and more accurately than traditional methods.

Personalized Medicine

By examining genetic, lifestyle, and health data, AI can tailor treatments to individual patients, improving results and reducing side effects.

Accelerated Drug Discovery

AI speeds up the development of new drugs by identifying promising compounds and predicting outcomes, cutting both time and cost in clinical trials.

2. Operational Efficiency and Patient Experience

Automated Administrative Tasks

AI handles routine functions like scheduling, billing, and record-keeping, freeing up staff to focus on care.

Resource Optimization

AI helps hospitals allocate beds, staff, and equipment more effectively by analyzing patient flow and usage patterns.

Virtual Assistants

AI-powered chatbots and apps support patients by answering questions, setting reminders, and scheduling appointments, enhancing access and convenience.

3. Advancing Research and Public Health

Faster Scientific Discoveries

AI can process massive datasets to uncover new insights, accelerating medical research.

Smarter Clinical Trials

AI helps design trials by selecting the best candidates and optimizing protocols for faster, more reliable results.

Disease Prediction and Modeling

Using epidemiological and patient data, AI models can forecast disease outbreaks and help identify at-risk populations for early intervention.

4. Ethical Considerations

Data Privacy and Security

AI systems depend on sensitive patient data, raising concerns about privacy. Strong protections are needed to prevent misuse.

Algorithmic Bias

If AI is trained on biased data, it can reinforce existing healthcare inequalities. Developers must ensure fairness and equity.

Transparency and Accountability

Complex AI systems often lack clear explanations for their decisions, making it difficult to assign responsibility for errors.

5. Microsoft Copilot's Perspective

Microsoft Copilot highlights AI's major roles in healthcare, including:

- **Medical Imaging & Diagnostics** – Detecting tumors, fractures, and abnormalities
- **Predictive Analytics** – Forecasting outbreaks and health risks
- **Personalized Medicine** – Tailoring treatments to genetics and history
- **Robotic Surgery** – Enhancing precision in minimally invasive procedures
- **Drug Development** – Identifying effective drug candidates faster

Q17. AI in Planning

AI planning is a subfield of artificial intelligence that focuses on helping systems create and execute a sequence of actions to achieve specific goals. These systems are designed to identify efficient paths from a starting point to a desired outcome using automated reasoning and planning algorithms.

How AI Planning Works

AI planning involves several key components:

- **Goal-Oriented Thinking** – AI planners are designed to achieve clearly defined objectives.
- **Action Sequencing** – The system selects and arranges the best series of steps to reach a goal.
- **Domain Models** – Planners use structured models of environments to understand available actions and their consequences.
- **State Space Search** – AI explores different possible scenarios or "states" to find the most effective route.
- **Domain Independence** – Some planners are adaptable to various fields without being tailored to a specific domain.

Key Applications

AI planning is used across many industries:

- **Robotics** – Planning movement and tasks for robots, such as warehouse navigation or product assembly

- **Logistics and Supply Chain** – Scheduling deliveries, optimizing transport routes, managing inventory
- **Workflow Management** – Automating complex processes in business, manufacturing, or service industries
- **Business Strategy** – Forecasting market trends, optimizing resources, and generating strategic plans

Benefits of AI Planning

- **Efficiency** – Automates complex planning, saving time and labor
- **Smarter Decisions** – Analyzes large datasets to select optimal plans
- **Adaptability** – Adjusts strategies based on changing data or constraints
- **Error Reduction** – Minimizes mistakes in high-stakes or repetitive planning scenarios

Microsoft Copilot's Perspective

Microsoft Copilot describes AI as a powerful force in modern planning. It enhances decision-making in areas like:

- **Project Management** – Automating schedules, streamlining workflows, flagging risks
- **Urban Planning** – Designing transport systems, managing public infrastructure, modeling environmental impact
- **Strategic Forecasting** – Predicting demand, identifying trends, and optimizing resource use

What sets AI apart is its ability not just to react, but to anticipate. AI planning tools offer insights that help planners stay ahead of challenges rather than simply respond to them.

Q18. General Intelligence as a Branch of Artificial Intelligence (AGI)

Artificial General Intelligence (AGI), also known as Strong AI, refers to a theoretical form of artificial intelligence capable of performing any intellectual task a human can. Unlike today's narrow AI systems—which are specialized for one task—AGI would possess the ability to reason, learn, and adapt across many domains, similar to human cognition.

Definition

AGI aims to create machines that can learn from experience, solve unfamiliar problems, and apply knowledge across different contexts without being retrained.

Narrow AI vs. General AI

- **Narrow AI** performs specific tasks (e.g., language translation, facial recognition).
- **General AI** would handle a broad spectrum of tasks with human-like versatility and understanding.

Core Characteristics of AGI

- **Human-like Cognitive Abilities** – Learning, reasoning, language understanding, and adaptability
- **Transferable Knowledge** – Ability to apply learning from one domain to another
- **Autonomous Problem-Solving** – Tackling new challenges independently

Theoretical Approaches

There are two leading schools of thought on how AGI might be developed:

- **Symbolic Approach** – Uses logic-based networks to mimic structured human reasoning
- **Emergentist Approach** – Suggests intelligence arises from complex interactions of simple units, similar to how neurons form cognition in the brain

Interdisciplinary Collaboration

Progress toward AGI requires input from:

- **Computer Science** – Algorithm design, data architecture
- **Neuroscience** – Understanding brain function
- **Cognitive Psychology** – Modeling learning and perception

Potential Benefits

If achieved, AGI could:

- Eliminate repetitive tasks and enhance productivity
- Drive innovation in science, healthcare, and education
- Allow humans to focus on creative, strategic, and relational roles

Current Status and Challenges

AGI remains a long-term research goal. Challenges include:

- Designing algorithms that generalize across tasks
- Ensuring alignment with human values and safety
- Understanding the nature of consciousness and reasoning

Timeline Expectations

Estimates vary widely. Some researchers believe AGI could emerge within a few decades; others argue it may require breakthroughs not yet imagined.

Microsoft Copilot's View

Microsoft Copilot frames AGI as AI's ultimate challenge—an intelligence that can reason, plan, adapt, and even demonstrate creativity. It warns that as AGI develops, ethical concerns must also grow: from ensuring AI safety to managing societal impact.

Whether through scaled-up current methods or entirely new paradigms, AGI continues to push the boundaries of what artificial intelligence can become.

Q19. Learning as a Branch of Artificial Intelligence

Learning is the ability to acquire new knowledge, skills, or behavior through experience, practice, or instruction. In humans, learning is a lifelong process that helps individuals adapt to their environment. In artificial intelligence, learning serves a similar purpose—it enables machines to improve over time and respond to new data.

Learning in AI

In AI, learning refers to how systems analyze data, recognize patterns, and adjust their behavior without being explicitly programmed for every scenario. This capacity allows AI to evolve beyond rigid rules and become more intelligent and adaptive.

Types of Learning in AI

- **Machine Learning** – AI learns from data by identifying patterns and making predictions or decisions based on past examples.
- **Supervised Learning** – The system is trained on labeled data, learning to predict outcomes based on known inputs and outputs.
- **Unsupervised Learning** – AI finds hidden patterns or groupings in unlabeled data.
- **Reinforcement Learning** – The system learns by receiving rewards or penalties, gradually improving its decisions.
- **Deep Learning** – A subset of machine learning that mimics the human brain using layers of artificial neurons to process complex data.

Why It Matters

Learning is what gives AI its power to grow, improve, and adapt:

- **Improves Accuracy** – Systems get better with more data

- **Enables Adaptability** – AI adjusts to new information in real time
- **Supports Complexity** – Learning allows AI to tackle tasks like speech recognition, image analysis, and language generation
- **Drives Innovation** – From medical diagnosis to personalized recommendations, AI learning is central to modern applications

Microsoft Copilot's View

Microsoft Copilot calls learning a foundational branch of AI. It allows systems to adapt, identify patterns, and become more useful over time. Machine learning and deep learning, in particular, help AI evolve past static instructions into systems that can reason, recommend, and even create.

Without learning, AI would simply execute rules. With learning, AI can think, predict, and improve.

Q20. AI in Education: A Specialized Branch of Artificial Intelligence

AI in education applies artificial intelligence to improve teaching, personalize learning, and support students and educators. From adaptive learning systems to virtual tutors, AI is reshaping classrooms and digital learning environments.

Key Applications of AI in Education

- **Personalized Learning**

  AI tailors lessons based on student performance, learning style, and pace—helping each learner master material in their own way.

- **Intelligent Tutoring Systems**

  Interactive AI tutors offer real-time feedback and adaptive explanations, supporting students as they work through new concepts.

- **Automated Grading and Feedback**

  AI speeds up grading of multiple-choice and short-answer assessments, giving students quicker feedback and freeing up teacher time.

- **Learning Analytics**

  AI tracks performance trends to help educators identify learning gaps, predict outcomes, and refine instruction.

- **Virtual Assistants and Chatbots**

  These tools support students with reminders, administrative questions, and even study guidance—available 24/7.

- **Immersive Learning Experiences**

  AI-powered virtual reality and augmented reality environments make learning interactive, visual, and engaging.

Benefits for Educators

- **Planning Support**

  AI can assist with lesson planning, generate differentiated materials, and suggest strategies to meet individual student needs.

- **Classroom Management**

  AI tools help teachers monitor student engagement and customize interventions.

Challenges and Ethical Considerations

- **Equity and Access**

  Not all students have access to AI-powered tools. Ensuring fairness and inclusivity is a major concern.

- **Data Privacy and Security**

  AI systems handle large amounts of personal data. Protecting that information is essential.

- **Teacher Training**

  Educators must learn how to use AI effectively. Professional development is key to successful integration.

- **Ethical Design**

  AI must be transparent, unbiased, and designed with student well-being in mind.

Future Outlook

AI's role in education will continue to grow. In the future, we can expect more sophisticated and responsive systems that enhance creativity, promote lifelong learning, and empower both students and teachers. The goal isn't to replace educators, but to amplify their impact.

Q21. Stages of AI as a Branch of Artificial Intelligence

Artificial Intelligence (AI) evolves through distinct stages, each marking an increase in complexity, capability, and autonomy. These stages help conceptualize how AI moves from task-specific systems to machines that might someday surpass human intelligence.

Core Stages of AI

1. Artificial Narrow Intelligence (ANI)

Also known as weak AI, ANI systems are designed for specific tasks. They include tools like voice assistants (Siri, Alexa), recommendation engines, and image recognition software. ANI can also be subdivided into cognitive models such as:

- **Reactive Machines** – No memory, just response to input
- **Limited Memory** – Learns from recent data
- **Theory of Mind (theoretical)** – Predicts emotions or intentions
- **Self-Awareness (theoretical)** – Possesses consciousness or identity

2. Artificial General Intelligence (AGI)

AGI, or strong AI, is still hypothetical. It envisions systems that can learn, reason, and solve problems across diverse domains, similar to a human. AGI would not need retraining to handle new tasks.

3. Artificial Superintelligence (ASI)

ASI refers to future AI systems that surpass human intelligence in every domain—logic, creativity, social interaction, and emotional awareness. It remains speculative and is often explored in philosophical and science fiction contexts.

Frameworks for AI Development and Adoption

Beyond capability stages, AI progress is also tracked using practical development models:

- **AI Project Cycle**

  Includes stages like project scoping, data acquisition, data exploration, modeling, and evaluation.

- **AI Adoption Stages**

  Describes how organizations move from awareness to experimentation, optimization, standardization, and finally, transformation through AI integration.

- **AI Maturity Models**

  Assess how deeply AI is embedded in an organization—from initial trials to enterprise-wide adoption.

- **AI Agent Evolution**

  Frameworks describe the shift from general-purpose AI chatbots to domain-specific experts and eventually fully autonomous AI-driven systems.

Microsoft Copilot's Summary

Microsoft Copilot outlines the AI stages as:

1. **ANI** – Specialized tools like chatbots and recommendation engines
2. **AGI** – Hypothetical systems with human-like reasoning and adaptability

3. **ASI** – AI that surpasses human intelligence in all areas

It notes that while ANI is widely used today, AGI and ASI remain aspirational, raising questions about feasibility, ethics, and impact.

To sum up, This chapter has shown that Artificial Intelligence is not a single tool or technology. It consists of several distinct branches, each with its own approach, purpose, and challenges.

The next chapter will move from structure to background. We will examine the history, goals, and developments that turned AI from a theoretical idea into a major force in modern technology.

The journey continues.

CHAPTER 2

FUNDAMENTALS OF MACHINE LEARNING

As we learned in chapter 1, machine learning enables computers to learn from data and make predictions without being explicitly programmed. It powers many tools introduced in Chapter 1, such as voice recognition and recommendation systems.

Q22. What is Machine Learning?

Machine learning is a branch of artificial intelligence that enables computers to learn from data and improve their performance over time without being explicitly programmed for each task.

Q23. What are the different types of Machine Learning?

Machine learning can be broadly categorized into three types: supervised learning, unsupervised learning, and reinforcement learning. Each type has different characteristics and is suited for different tasks.

Supervised learning involves training a model on a labeled dataset, where each input has a corresponding output. The model learns to map inputs to outputs, allowing it to make predictions on new, unseen data. Common applications include email spam detection and image classification.

Unsupervised learning works with unlabeled data. The model identifies patterns, groupings, or structures within the data without specific outputs provided during training. Examples include customer segmentation and anomaly detection.

Reinforcement learning is based on the idea of agents learning by interacting with an environment. The agent receives feedback in the form of rewards or penalties and uses this information to learn a strategy for achieving its goals. This type of learning is often used in robotics and game playing.

Q24. What are the Key Components of Machine Learning?

Now that we've covered AI's broader development, let's take a closer look at how machine learning systems are actually built and used.

1. **Data**: High-quality and relevant data is the foundation of any machine learning system. The data used for training must be representative of the problems the model will encounter in real use.
2. **Features**: Features are the individual measurable properties of the data that the model uses to learn patterns. Feature selection and engineering are critical steps that affect the model's performance.
3. **Model**: A model is the mathematical representation that maps input features to output predictions. Different algorithms produce different models, such as decision trees, support vector machines, or neural networks.
4. **Training**: Training involves feeding data into the model so it can learn patterns and relationships. The model adjusts its internal parameters to minimize errors between predicted and actual outcomes.
5. **Evaluation**: After training, the model must be evaluated to ensure it performs well. Evaluation metrics, such as accuracy or precision, help determine how well the model generalizes to new data.
6. **Prediction**: Once trained and evaluated, the model can be used to make predictions on new input data. This is the stage where machine learning becomes practical for real-world use.

Machine learning is a foundation of artificial intelligence. It allows systems to learn from data and improve over time. With a basic understanding of these principles, we can begin to explore how machine learning is applied in practical ways.

CHAPTER 3

NEURAL NETWORKS AND DEEP LEARNING

Neural networks are central to deep learning, a subfield of machine learning. They are designed to recognize patterns in data by mimicking the structure and function of the human brain. Deep learning expands on this concept by using neural networks with multiple hidden layers, allowing systems to identify complex patterns and relationships in large datasets.

A neural network is like a giant team of decision-makers passing notes. Each layer processes the information a little differently, building on what came before. By the time the final layer speaks, it has a well-informed answer based on the collective "thinking" of all the layers before it.

Q25. What Are Neural Networks?

Neural networks are computational models made up of layers of interconnected units called neurons. Each neuron receives input, processes it using a mathematical function, and passes the result to the next layer. These layers typically include an input layer, one or more hidden layers, and an output layer.

Q26. Understanding Deep Learning

Deep learning uses neural networks with more than three layers to learn from data. The term "deep" refers to the number of hidden layers in the model. Each layer helps the system extract increasingly abstract features from the input, making it capable of understanding complex patterns such as images, language, or sound.

Q27. How Neural Networks Learn

Neural networks learn through a process called training. During training, the system is given input data along with the expected output.

It makes predictions and compares them with the correct answers. The difference between the prediction and the actual result is called an error. Using an algorithm known as backpropagation, the network adjusts its internal parameters to reduce this error over time.

Q28. Applications of Deep Learning

Neural networks and deep learning are used in various fields. In healthcare, they assist with medical imaging and diagnostics. In finance, they detect fraudulent transactions. In consumer technology, they power voice assistants, facial recognition tools, and recommendation systems. These systems improve performance as they are exposed to more data.

Q29. Challenges and Limitations

Deep learning requires large amounts of data and significant computing power. It also raises challenges in interpretability. The decisions made by deep networks can be difficult to understand, leading to what is often called the "black box" problem. This lack of transparency can make it hard to trust or explain certain outcomes.

To sum up, neural networks and deep learning play a foundational role in modern artificial intelligence. They allow systems to learn from data, recognize complex patterns, and perform tasks that once required human expertise. While powerful, they also bring challenges related to data requirements, computational cost, and interpretability. Understanding these models provides insight into how AI systems analyze and learn from data.

CHAPTER 4

Natural Language Processing

When you speak to a voice assistant, get grammar suggestions, or translate languages online, you are using NLP. It powers systems that analyze text, understand meaning, and produce language that makes sense. This chapter introduces NLP, its key tasks, and its growing impact in areas such as communication, business, and education.

Q30. What is NLP?

Natural Language Processing (NLP) is a branch of artificial intelligence that helps machines understand and respond to human language. It combines computer science and linguistics to make interactions between humans and computers more natural. Just as a child learns language by listening, reading, and speaking, NLP allows machines to perform similar tasks with words and text.

Q31. Core Tasks in NLP

NLP includes a wide range of tasks that help computers understand, interpret, and generate human language. Some of the most common tasks are:

1. **Tokenization**: Breaking down text into individual words or phrases. For example, the sentence "The cat sat on the mat" becomes ["The", "cat", "sat", "on", "the", "mat"].

2. **Part-of-Speech Tagging**: Identifying the role of each word in a sentence, such as noun, verb, or adjective.

3. **Named Entity Recognition (NER)**: Finding and classifying names of people, organizations, places, and other specific items.

4. **Sentiment Analysis**: Determining whether a sentence or document expresses a positive, negative, or neutral opinion.
5. **Machine Translation**: Translating text from one language to another, such as English to French.
6. **Text Summarization**: Producing a shorter version of a longer text while preserving key information.
7. **Speech Recognition**: Converting spoken language into text.
8. **Text Generation**: Creating readable, coherent text based on input or prompts.

Q32. Approaches in NLP

There are two major approaches to building NLP systems:

1. **Rule-Based Systems**: These systems use predefined rules written by experts. They are often accurate for specific tasks but can be hard to scale.
2. **Statistical and Machine Learning Models**: These systems learn from large datasets. They adapt better to new tasks and languages and are more flexible than rule-based systems.

Recent NLP systems use deep learning techniques and large-scale neural networks. These models, such as transformers, are trained on billions of words and can perform multiple NLP tasks with high accuracy.

Q33. Applications of NLP

NLP is used in many areas of everyday life and professional settings:

- **Virtual Assistants**: Voice-controlled assistants rely on NLP to understand and respond to commands.
- **Search Engines**: NLP improves search results by understanding the intent behind queries.
- **Email Filtering**: Spam detection systems use NLP to identify unwanted messages.

- **Translation Services**: Apps and platforms translate languages using NLP.
- **Customer Support**: Chatbots answer questions using natural language.
- **Healthcare**: NLP helps extract useful information from clinical notes and medical documents.

As AI continues to grow, NLP will remain a key area of development, opening new possibilities in education, health, business, and beyond.

CHAPTER 5

AI in Real-World Applications

Imagine trying to manage a city without traffic lights, street signs, or maps. Everything would depend on instinct, memory, and guesswork.

AI in today's world acts like a system of signals and guides that help us navigate information, decisions, and processes more efficiently.

Just as a well-designed infrastructure brings order to movement, AI brings structure and intelligence to digital tasks that would otherwise be overwhelming or chaotic

Artificial Intelligence has moved from labs and theory into daily routines. It now powers countless systems, tools, and services across industries.

Understanding where and how AI is applied helps clarify its value and the ways it is shaping human interaction with technology.

Q34. Personal and Consumer Applications

AI is built into devices people use every day. Voice assistants such as Siri, Alexa, and Google Assistant help with scheduling, reminders, and quick information searches. Streaming platforms use AI to suggest shows or songs based on past choices.

Social media uses AI to organize content feeds, recognize images, and detect inappropriate posts. In online shopping, AI systems recommend products, assist with virtual try-ons, and provide customer service through chatbots.

One example is Spotify's recommendation engine. It uses machine learning models trained on listening habits, genre similarities, and community behavior to curate personalized playlists like Discover Weekly.

These suggestions evolve as user preferences change, often increasing engagement and music discovery.

Q35. Healthcare and Medical Innovation

Healthcare is one of the most rapidly growing fields for AI. Systems trained on medical images can help detect diseases such as cancer at early stages.

AI is also used to sort patient records, manage hospital workflows, and assist in robotic surgeries. In drug development, AI models identify possible compounds and simulate how they might behave, speeding up the research process. Personalized medicine benefits from AI by tailoring treatments based on an individual's genetics and health data.

An example is the use of AI at Moorfields Eye Hospital in London. In collaboration with DeepMind, they developed an AI system that scans retinal images to detect over fifty eye diseases as accurately as expert doctors. This early diagnosis allows for timely intervention and preserves patients' vision.

Q36. Finance, Retail, and Manufacturing

In finance, AI supports fraud detection by spotting irregular patterns in transactions. It evaluates credit risk, automates trading strategies, and offers personalized investment advice. Retail uses AI for inventory control, personalized marketing, and sales forecasting.

In manufacturing, AI improves production efficiency through predictive maintenance, where machines are repaired before failure occurs. Quality control is also automated using computer vision systems.

An in-depth example comes from Amazon's supply chain. AI forecasts demand across locations using real-time purchasing behavior and market trends. This prediction helps pre-stock items in nearby fulfillment centers, ensuring faster delivery and reducing shipping costs.

Q37. Cybersecurity and Risk Management

AI tools in cybersecurity monitor network traffic, detect threats, and isolate vulnerabilities. These systems respond faster than humans to potential breaches.

In risk management, AI evaluates market shifts, detects insurance fraud, and helps organizations prepare for economic uncertainty.

A concrete example is Darktrace, a cybersecurity company that uses AI to create a model of normal network behavior. When deviations occur, such as data transfers to unknown servers, the system flags them immediately and can neutralize the threat before human analysts respond.

Q38. Emerging Frontiers: Robotics, Self-Driving Cars, and Generative AI

Robotics, guided by AI, is expanding into areas like warehouse automation, precision farming, and elder care. Self-driving vehicles rely on AI to process real-time sensor data and make navigation decisions.

Although full autonomy remains under development, major advances continue in traffic recognition and route optimization. Generative AI allows machines to create new content such as text, images, audio, and code. This is changing how professionals design marketing campaigns, write drafts, or develop visual assets.

One advanced example is the AI used in Tesla's Autopilot. It continuously gathers data from millions of driving scenarios and uses neural networks to improve decision-making. This includes detecting pedestrians, reacting to traffic lights, and adjusting driving behavior based on the flow of traffic.

AI has become a central part of modern life. But as it continues to evolve, it is important to ask who benefits, how decisions are made, and how ethical concerns are addressed.

CHAPTER 6

ETHICAL CHALLENGES IN AI

Artificial Intelligence presents major ethical and societal challenges. These include concerns about bias, privacy, security, accountability, job displacement, and the potential misuse of AI technologies. Addressing these issues requires collaboration across disciplines, including technology, law, philosophy, and policy-making.

Q39. Bias and Discrimination

AI systems trained on biased datasets can produce unfair outcomes. This is especially problematic in high-stakes areas such as hiring, lending, or law enforcement.

For example, in hiring software, if the training data reflects a history of gender bias, the AI may unfairly favor male candidates. For instance, a resume-screening tool once penalized resumes with terms like "women's chess club captain," leading to discriminatory outcomes.

Q40. Data Privacy

AI depends on large datasets, many of which contain personal information. This raises concerns about data collection, storage, use, and misuse.

In health care, for instance, AI models trained on patient records can improve diagnosis but also expose sensitive personal information if data security is weak. A breach can lead to identity theft or unauthorized use of health data.

Q41. Accountability and Transparency

AI systems often make decisions that are difficult to interpret. This raises important questions: Who is responsible when an AI system makes a harmful decision?

For instance, if a self-driving car causes an accident, it is unclear whether the blame lies with the manufacturer, the software developer, or the user.

Q42. Job Displacement

As AI automates tasks, concerns about employment grow. Some jobs may be transformed, while others may disappear entirely.

In the financial industry, for example, AI is automating tasks like risk analysis and fraud detection. While efficiency increases, entry-level analyst positions are shrinking, affecting career entry points.

Q43. Surveillance and Misuse

AI technologies can be used for mass surveillance or by authoritarian regimes to track citizens.

Facial recognition systems are deployed in some cities to monitor public gatherings, raising alarm over the erosion of civil liberties and misuse by law enforcement.

Q44. Existential Risks

Advanced AI systems may eventually pose existential risks, whether through unintended behavior or malicious use.

If a highly autonomous AI were to pursue goals that conflict with human values and there is no fail-safe mechanism, it could cause harm on a wide scale. This is an example of an AI existential risk.

Q45. Addressing the Ethical Concerns

- **Explainable AI (XAI):** Creating models that provide understandable reasoning behind decisions helps build trust and ensures accountability.
- **Bias Mitigation:** Techniques such as data auditing, fairness-aware algorithms, and inclusive training datasets aim to reduce discrimination.
- **Data Protection:** Laws like the GDPR set standards for how personal data can be collected and used.

- **AI Governance:** Governments and organizations are creating ethical guidelines and governance frameworks.
- **Public Education:** Increasing awareness and education about AI empowers citizens to engage with and shape ethical standards.

To sum up, ethical challenges in AI are serious and wide-ranging. A coordinated approach involving ethical frameworks, public accountability, transparent systems, and inclusive policy-making is essential to ensure AI serves humanity positively.

CHAPTER 7

THE FUTURE OF AI

As mentioned in Chapter 5, AI is steadily becoming a core part of daily life. In the near future, AI will no longer be seen as a technical tool used only by specialists. Instead, it will function as a collaborative partner in tasks that demand precision, personalization, and adaptability.

This shift will increase productivity, raise new ethical challenges, and change the skills needed for future jobs.

Q46. Human-AI Collaboration

AI will not replace human thinking but will support it. People will rely on AI systems to process data quickly and provide insights that guide better decisions.

In architecture, for instance, designers will use AI to simulate how a building reacts to different weather patterns, population flows, and materials. The AI will not design the building but will run millions of scenarios, allowing the architect to make more informed choices.

Q47. Personalized Experiences

AI will deliver tailored experiences in education, healthcare, and finance by learning individual preferences and needs.

For example, in healthcare, AI will monitor a patient's health using wearable devices and predict issues before they arise. For example, it may alert doctors to subtle changes in a heart patient's routine that could signal an early risk of arrhythmia.

Q48. Autonomous Systems

Machines will increasingly act without human input, performing actions based on their training and environmental data.

Autonomous drones will be used to inspect damaged infrastructure in disaster zones. They'll navigate dangerous areas without putting humans at risk and send real-time updates to emergency teams.

Q49. Efficiency and Automation

Repetitive and time-consuming tasks will be handled by AI, freeing humans to focus on more complex work. E.g., in logistics, AI will optimize delivery routes by analyzing traffic, fuel costs, and inventory needs. This can lead to faster deliveries and reduced operational waste for shipping companies.

Q50. Ethical and Social Challenges

As mentioned in the previous chapter, with more AI comes more responsibility. Developers and regulators will need to address concerns about bias, misuse, and accountability.

An example can be an AI hiring tool that might inadvertently favoring certain resumes over others. Companies will need auditors who specialize in identifying and correcting unfair algorithms to ensure fair hiring practices.

CONCLUDING THOUGHTS ON ARTIFICIAL INTELLIGENCE

Artificial Intelligence is no longer a distant concept. It now shapes the tools we use, the systems we depend on, and the way we interact with information and each other. From simple pattern recognition to the complexities of neural networks, from chatbots to autonomous vehicles, AI has moved steadily from theory into everyday practice.

This book has introduced core concepts like machine learning, natural language processing, and deep learning. It has explored practical applications in healthcare, education, business, and art. It has also raised important questions about ethics, explainability, and the future role of humans in a world increasingly influenced by intelligent machines.

AI continues to evolve. New models will emerge. More data will be gathered. Capabilities will expand, but so will the responsibilities. Policymakers, developers, educators, and users must all remain alert to how these tools are built and used. Trust in AI will depend not only on its technical performance but also on fairness, transparency, and accountability.

The journey ahead is uncertain. AGI may remain theoretical for some time. But narrow AI, when used well, can solve real problems, extend human capacity, and support better decision-making across fields. Learning to understand and question AI systems is now a basic skill, not a luxury.

This is not the end of the AI story. It is the beginning of a conversation. A responsible, informed public will shape how this powerful technology unfolds.

APPENDIX A

DATA SOURCES TO TRAIN AI MODELS

High-quality and diverse data is essential for training AI models, especially generative models that create text, images, videos, or audio. Below is an organized overview of key data sources and methods used across AI domains.

1. Public Datasets

Public datasets are open and accessible for academic and commercial AI training. These include:

Text and Language Models

- **Common Crawl** – A massive repository of web-crawled text data.
- **Wikipedia** – Rich in structured and unstructured textual content, ideal for language modeling and knowledge extraction.
- **LibriSpeech** – A corpus of read English speech, used in training speech recognition systems.

Images and Videos

- **ImageNet** – Contains millions of labeled images, widely used for object recognition.
- **COCO (Common Objects in Context)** – Used for object detection and captioning.
- **MSR-VTT** – A benchmark dataset for video-to-text tasks.

Other Sources

- **Kaggle** – Offers a wide array of datasets in different formats, suitable for machine learning competitions and training projects.

- **Data.gov and Google Public Data** – Include structured datasets across healthcare, demographics, transportation, and economics.
- **Google Dataset Search** – Helps locate datasets across domains.
- **Open Images and MNIST** – Serve as standard datasets for object detection and handwritten digit recognition, respectively.

2. Crowdsourced and Curated Data

Crowdsourcing Platforms

- **Clickworker** and **Appen** – Connect you to a global workforce that can generate, label, and verify data for training models.
- **Humans in the Loop** – Specializes in human-verified training datasets for computer vision and AI ethics projects.

Data Vendors and Labeling Services:

- **SmartOne.ai**, **Innodata**, **Label Studio**, **Aya Data**, **Macgence** – Provide domain-specific labeled datasets and annotation services.

3. Internal and Synthetic Data

Internal Data

- Businesses often use proprietary datasets such as customer transactions, service logs, and sensor outputs to train custom AI models tailored to their goals.

Synthetic Data

- Generated using simulations or data augmentation techniques, synthetic data is useful when real-world data is scarce, expensive, or sensitive.

4. Data Collection Techniques

Web Scraping

- This method involves extracting content from websites automatically, useful for niche or up-to-date data needs. It requires compliance with data use policies and ethical guidelines.

User Interaction Logs

- When permitted, user behavior data can be used to train models for personalization, recommendations, or adaptive interfaces.

APPENDIX B

EXPLORING AI HARDWARE, EXPLAINABILITY, AND CREATIVITY

1. AI Hardware: Powering Intelligence

AI relies not only on algorithms but also on powerful hardware that makes computation feasible. Specialized components like GPUs and TPUs process vast amounts of data in parallel, accelerating training and inference. General-purpose CPUs still serve vital roles, especially in less demanding AI applications.

2. Explainability: Making AI Understandable

As AI systems become more embedded in everyday life, understanding how they make decisions is critical. Explainability provides transparency. It allows developers and users to see which features influenced a decision, building trust and supporting accountability.

3. AI in Art and Creativity: New Frontiers of Expression

AI is reshaping art, music, and design. Tools like DALL·E and Midjourney generate artwork from text prompts. Other models write poetry, compose symphonies, or suggest new design layouts.

These topics—hardware, explainability, and creativity—demonstrate that AI is not just about logic and prediction. It also touches infrastructure, ethics, and imagination. Understanding these aspects enriches our view of AI's role in society and its future possibilities.

ABOUT THE AUTHOR

Hirendra Majumdar, M. ASCE, PE, is a highly accomplished civil engineer with a career spanning over five decades. His professional journey is rooted in water resources, hydraulics, and civil engineering, with a wealth of experiences across state agencies, private firms, and international projects. Holding both a master's degree from the University of Wisconsin, Madison (1964) and a bachelor's degree from the prestigious Indian Institute of Technology, Kharagpur (1960). Mr. Majumdar is a registered Professional Engineer in California.